A TELEOLOGICAL DESCRIPTION OF THE UNIVERSE

CHARLES A. CUMMINGS

This book is dedicated to my wife, children, grandchildren, and all who look into infinity with ever-increasing speed and insight; not only with eyes but spirit that is not confined in time or space.

About the Author

Charles A. Cummings received a degree in Electrical Engineering from Franklin University in Columbus, Ohio in 1960. His studies continued at the University of Cincinnati, where he received a degree in Electro-Mechanical Engineering in 1975. Over the last 50 years Charles took graduate courses primarily in engineering. Many focused on the field theory. These discoveries are still relevant today. .

The developmental activities in recent years, including the discovery of instantaneous information transfer (intergalactic) that do not appear to be part of electromagnetic fields, have driven these quantum embedded studies.

Author's Note

In this text the past, present, and future are all one, immutable. Everything that is existed in the past, present, and through eternity. Here named "The Dynamic Vector," which everything is created from: galaxies, solar systems and life.

Section I
THE EVIDENCE: MATERIAL FORMS

The universe consists of one particle; its dynamics create everything. An expanding universe must consist of a dynamic source, here described as a Dynamic Vector. The Dynamic Vectors are the reason the universe exists.

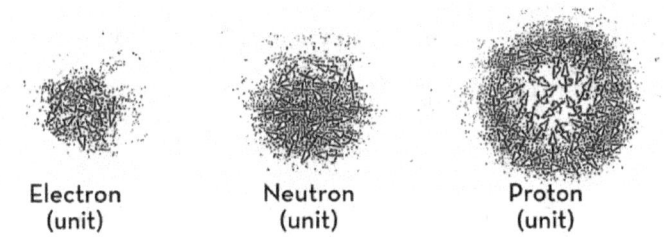

| Electron (unit) | Neutron (unit) | Proton (unit) |

Figure 1: Examples of the Dynamic Vectors making up a Few Components of the Universe

The Dynamic Vectors are greater in number than that detected or conceived. If one stands and looks at the entire universe, everything (matter and energy), it may then be determined that everything consists of the same constituents.

Consider the momentum p of photons. The quantum theory states that their energy is proportional to the frequency $E = h\nu$. Here h denotes Planck's elementary quantum action; therefore, the energy contained in the Dynamic Vector may be:

$E = h\nu$ the lowest energy constituent of any particle.

Where: h = Planck's constant – 6.625×10^{-34} *joule.sec*

v = frequency

c = speed of light

E = energy

The relativistic energy momentum for a free particle being an intricate particle of the universe has no potential energy. The energy of the subject "particle," the Dynamic Vector, is defined above.

The orientation of each vector direction and spin is fixed in place defining a particle. Therefore, every electron, as well as every other unit of mass or energy, is defined by the fixed orientation of the vectors defining each particle. Could this defined universal structure of the universe be too complex? Consider all the theories and discoveries that have surfaced in the last 100 years: mass-energy $E = mc^2$. string theory, dark matter, black holes, and everything may possibly be explained by the Dynamic Vector. The 100+ elements that appear in the Periodic Table may be a good starting point to consider what may be fundamental to the universe.

The electron, neutron, and proton are only examples of three "units" that make up matter as we know it as illustrated in Figure 1. Every part of the universe may be constructed in a similar manner.

Entropy may be controlled by the Dynamic Vector. As the entropy decreases, everything increases in order, down to life and the intricacy of replication, cell division directed by the Dynamic Vector, a complex control, creating living organisms and their replication.

The entity that makes up everything must be indivisible. Relative orientation and replication, magnitude, and spin may be beyond quantitative definition.

In terms of a fixed *frame of reference*, a tensor can be represented as an organized *multidimensional array*. The order of a tensor is the dimensionality of the array needed to represent the number of indices needed to label a component of that array. A linear map is represented by a matrix, a two-dimensional array and therefore is a 2nd-order tensor. The linear mapping may be extended here to a three-dimensional array creating a 3rd-order tensor. Such relationships include products, cross products and mapping the Euclidean vectors. Scalars are single numbers and are thus 0th-order tensors. Because they express a relationship between vectors, tensors themselves must be *independent of a particular choice* of a *coordinate system.* The coordinate is independent of a tensor that takes the form of a "covariant" transformation that relates the array computed in one coordinate system of that computed in another one. The tensor type is a pair of natural numbers (*n, m*), where *n* is the number of contravariant indices and *m* is the number of covariant indices. In this case the total order of a tensor is the sum of these two numbers.

An analytic expression defining the universe may not be possible; however, the origin of everything making up the universe could be established by considering what may be necessary to allow all the parts, cell growth and the interaction of the cells, including replication exists. The development of the entire universe, solar systems to galaxies and the interrelationship down to the simplest forms of life, may yield a clue.

Likely the most difficult question is: What method is available that will detect and measure the constituents of the universe as proposed? "The singularity" – the dynamic organization of the constituents (magnitude, spin and direction) may be the essence of the universe.

The Dynamic Vectors, being the smallest of all units, may not be seen; however, they could be detected by the virtue of their orientation to each other and density. The cloud of vectors could appear as energy within the environment of matter we see around us. It is well expected that the universe is over 90% dark energy and 5% dark matter. What we see (about 3%) is therefore orientation of vectors comprising particles such as electrons, neutrons and protons that may be detected by the Dynamic Vector itself.

Section II
THE EVIDENCE: LIFE FORMS

The orientation, direction and spin within a cluster (the number in a cluster is determined by each Dynamic Vector) determines the cluster's functionality. The Dynamic Vector knows and controls all that has been and that is being clustered. All things: past, present and future are eternal. The vectors in Figure 2 are tensors: 1,2… all the same in control and functionality. Their spin rotations determine the dimension they are in.

Figure 2: *A cluster of three Dynamic Vectors (a cluster may contain one vector to as many required to establish a specific functionality).*

Over the years, life will be making a transition from what we observe (living cells) to forms that require no atomic structure. Consequently, life may be a "field," possibly a form of energy not observed or exhibited within the electromagnetic fields discovered over 200 years ago.

It may be highly probable that a message transfer wave "field" is this energy (action at distance everywhere simultaneously), and may control and keep everything together. Therefore, life may not be atomic; consequently, there would be no deterioration with time. Entropy, a measure of ordering, has been accepted as a law by most scientists. Evidence is: there is no system in the universe that will order itself. A system will remain the same in function or deteriorate with time. Possible ordering is not internal but external, a message transfer wave or energy that may be infinite in control and proliferation.

A revelation of this information was revealed over 60 years ago when the "design" of the DNA (double helix) was discovered that led to the understanding that the gene is the recipe for how to make protein. This information is "written" on the long DNA molecule designated as A, G, C, and T (letters chosen by the DNA discoverer) which are copied by an enzyme making up

14

the RNA, the message carrier. This extremely complex ordering is far beyond chance or evolution. This DNA information is included in this discussion to support the argument for the Dynamic Vector, which is argued to be eternal.

Order being eternally established may mean that life on earth, physically made up of electrons, protons, and neutrons which constitute atoms that make up over 100 elements, are part of the composition of life. That is, these elements are observed happenings making up earth, as well as everything in the universe. The energy and message transfer waves comprise everything.

The message transfer happening within the physical, atomic structure is eternal. Elements are left behind leaving only the infinite message transfer waves that have made up our existence from the time of inception.

Life does not start and end, but begins and grows physically in form field or energy! When transformation comes, the physical is left behind and the energy that has always been the life, continues eternally.

Life may not be physical as we have learned through

knowledge of the DNA, but a "spirit" that does not require the physical components so real to man's knowledge of existence.

The components that make up life may also be the components that make up the universe but dynamic in form and replication. These components are electrons, protons, and neutrons (e, p, n). All that one touches by hand or mind, from the one celled amoeba to the universe, are comprised of these three parts including recent findings referred to earlier. The reason for generating the list above is to emphasize the finite part count (e, p, n) that comprises the universe.

A question that may exist is: how do the electrons, protons, and neutrons form life, including replication? Could the underlying control be ubiquitous, the three-part structure (e, p, n) being entirely under "its" control? The underlying structure of the universe is made up of atoms (over 100), all of which consist of e, p, and n. From the time of the big bang (widely accepted but possibly questionable) that started over thirteen billion years ago, electrons, protons, and neutrons were formed. When did "life" start and how may be an open-ended question.

If one accepts that the expansion and formations of the system called the universe are under the Dynamic Vector's control, then it may be logical to accept further control.

These systems divide, replicating themselves while forming different, unique parts that make up a living organism. Further, there must be communications between all cells to ensure a reliable functional operation and replication. Going further, the developed operation called life includes cells that carry reproductive information by the host, some internally and externally creating plant life and living, moving, functional creatures, male and female. When the (at least) two pieces of information are brought in contact with each other, a new life form of the host species is created.

Further: Plant life replicates itself through an embryo in a protective coat that is activated in the presence of water and soil. The plant generates a seed containing all the information to replicate itself, starting a duplicate of the host plant.

Possibly a difference exists among the many forms of animal and plant life that are outside the replications covered in the previous discussions. Another mysterious eternal control may be the transformation that a life encounters when body function ceases, but the energy defining the being continues. This functionality, the extent being more fully covered above, may be in all living organisms and is the essence of life.

Section III
THE EVIDENCE: INTEGRATION OF SECTION I AND II

As far as I can historically verify, Maxwell was the first to decide that physical imagination is of greater value than a competence in handling symbols. In fact, he actually published this concept in a large treatise: "...I perceive that Faraday's method of conceiving the phenomena was also a mathematical one, though not exhibited in the conventional form of mathematical symbols. Therefore, it is possible to think mathematically without knowing mathematics, just as it is possible to appreciate and even compose music without having had musical training." Einstein later paraphrased this same idea when he said, "Imagination is more important than knowledge."

This communication is the Dynamic Vector (all inclusive) creating intelligent transfer of knowledge. The workings of the physical world may only appear to be complex due to our lack of basic concepts. The high degree of mathematical involvement even when analysis is kept linear requires a large number of symbols, i.e., definitions to adequately define and solve many classes of problems. Of course, the non-linear analysis required due to the accuracy requirements of some problems is not only more involved in symbolism and definitions, but many times requires "logical assumptions."

Possibly the physical world is not non-linear; analysts are! Of course, the possible reason analysts are non-linear is a result of their vantage point in space and time. Analysts are on the inside looking out, but with greatly impaired vision. The place to "watch the show" is outside, to include "all" temporal and spatial events.

It is not likely that a "man" will ever occupy the "outside seat," but as more information is accumulated a "point" will be reached where analysis will begin to become simpler in symbolism and definition and will likely continue to increase for some time to come. As man delves into a physical world with ever-increasing speed of "insight" his means of understanding becomes more complex. It is now apparent, to me, that the means of understanding or an "outside" understanding is imperative to keep from getting bogged down in the mire of man's own symbolism and definitions.

It may be concluded that the teleological nature of the universe will never be completely revealed since man is finite and does not appear to have the capacity to understand the infinite. How may the degree of knowledge be measured when the extent of man's knowledge is relative and therefore has no quantitative significance when viewed on an infinite scale?

Summary

The presence of God (divine creator of the universe) is not mentioned in the text. The descriptions of the source of elements, particles, and life dynamics have been covered in Sections I, II and III. All descriptions support the eternal nature of the universe that: God and the Dynamic Vector are synonymous.

Past – Present – Future (the firmament being the same Dynamic Vector eternally). All the "vectors" throughout the universe are in continuous contact with every other "vector" throughout the universe.

We have just begun the journey.